Grandfather's Nose
Why We Look Alike or Different

by Dorothy Hinshaw Patent

illustrations by Diane Palmisciano

Franklin Watts
New York London Toronto Sydney
1989

For the teachers who gave me a love of genetics
D. H. P.

For my new friend Nancy P.
D. P.

Library of Congress Cataloging-in-Publication Data

Patent, Dorothy Hinshaw.
Grandfather's nose.

(A Discovering science book)
Summary: Discusses basic genetics, explaining how the combination of genes passed on from our parents makes each of us a unique individual.
1. Genetics—Juvenile literature. [1. Genetics]
I. Palmisciano, Diane, ill. II. Title. III. Series.
QH437.5.P37 1989 575 89-9140
ISBN 0-531-10716-7

Printed in the United States of America
6 5 4 3 2 1

Looking Alike, Looking Different

Do you look like your mother or father? Do people mix you up with your brother or sister? People usually do look like their relatives in some ways. They have a "family resemblance."

Other animals, and plants, too, have family resemblances. Two German shepherd parents have German shepherd puppies, but two poodles will have baby poodles. Robins look like robins, and rats are clearly rats.

How is a grandfather's nose or a mother's red hair passed along to a daughter or grandson? That is, how are these features inherited? When you inherit something, you receive it from your parents. Scientists try to understand how we inherit family traits, or characteristics. They call their work *genetics* (jen-NET- icks).

For a long time, people had thought that traits from the father and mother were mixed together in their children. They saw that the skin color of children, for example, can look like a mixture of their parents' color.

But many traits don't look blended or mixed together. A blond father and a black-haired mother do not usually have children with brown hair. They will have either blond or black hair.

The Father of Genetics

Gregor Mendel was a monk who lived in Austria. He wanted to understand why some plants and animals looked alike and others looked different.

Mendel discovered that family traits are somehow passed from parents to their children as if the traits were particles, or tiny packages, that couldn't be divided or destroyed.

Mendel's Experiments

Gregor Mendel decided to study pea plants. He could grow peas easily in the garden of the monastery where he lived.

He carefully chose a group of tall pea plants and another group of short pea plants. Then he *crossed,* or mated, the tall and short pea plants with each other. He did this by taking some pollen from the male part of the flower of one plant. He then put the pollen on the female part of the flower of the other plant. He found that all the baby plants were tall. It looked as if the short trait had disappeared. But then, if he crossed these baby plants with each other, about one out of every four of the new plants was short! The short-plant trait had come back.

The same thing happened when he crossed a violet-flowered plant with a white-flowered one. All the new plants had violet flowers. But if those new violet plants were crossed with each other, about one-fourth of the new plants had white flowers.

Mendel decided that each plant carried two particles of the material that was responsible for each trait. We call these particles, or bits, *genes* (just like the name JEAN). Each gene can occur in more than one form—the way ice cream comes in different flavors. These different forms are called *alleles* (a-LEELS). Mendel discovered that the gene for height in peas has two alleles, a tall allele and a short one. The flower color gene in peas has at least two alleles, a violet one and a white one.

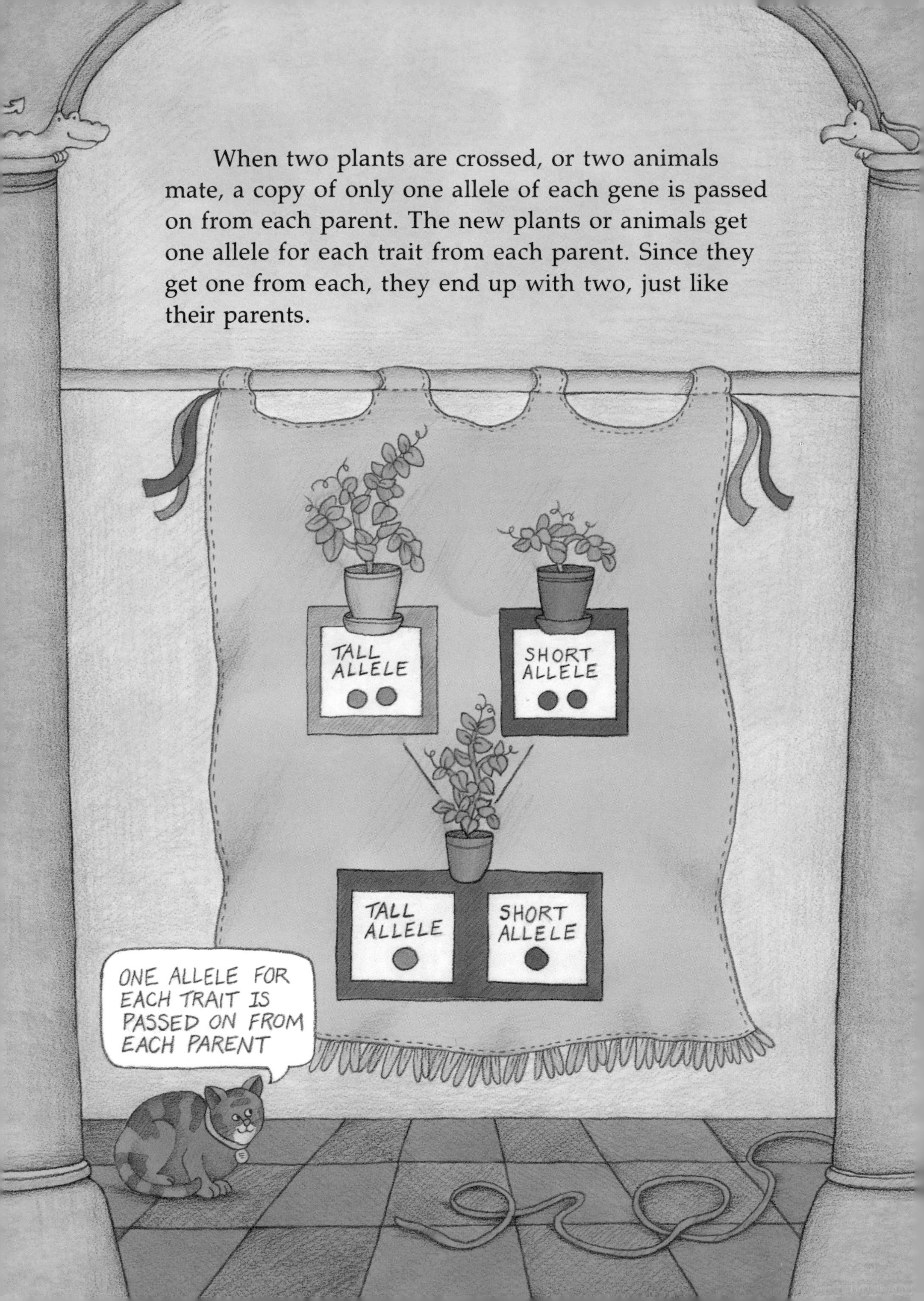

When two plants are crossed, or two animals mate, a copy of only one allele of each gene is passed on from each parent. The new plants or animals get one allele for each trait from each parent. Since they get one from each, they end up with two, just like their parents.

The Meaning of Dominant

Why did only one allele of the gene show itself when tall and short plants were crossed? Some traits are *dominant* (DOHM-i-nahnt) over others. Dominant means strong or powerful. The other allele is called *recessive* (ri-SES-iv). Recessive means weak. The recessive allele won't show up if it is paired with the dominant one. But it is still there. It just doesn't appear when the dominant one is present.

When Mendel crossed violet- and white-flowered peas, he started with a plant carrying two violet-flower alleles. He crossed it with a plant that had two alleles for white flowers. Each of the new plants got a violet allele from one parent. It also received a white one from the other parent. Violet is dominant over white, so all the new plants had violet flowers.

Although the new plants looked the same as their violet-colored parent, they were different. Each plant had one violet allele and one white one. When these plants were crossed, about half of the next generation got a violet allele from the female parent. The other half got a white one. Half received a white allele from the male parent, and half got a violet one.

If a plant got a violet allele from one or both parents, it had violet flowers. But if it got a white allele from *each* parent, it had white flowers, just like one of the grandparents. The white flower trait didn't show up when paired with the violet one. But it didn't change or disappear either. It was there all the time.

Where Are the Genes?

Most living things are made up of thousands and thousands of tiny cells. Each cell contains stringlike threads called *chromosomes* (KROHM-oh-sohmes). The chromosomes are in the center part of the cell, called the nucleus (NOO-klee-uhs). The genes are lined up on the chromosomes, like beads on a string.

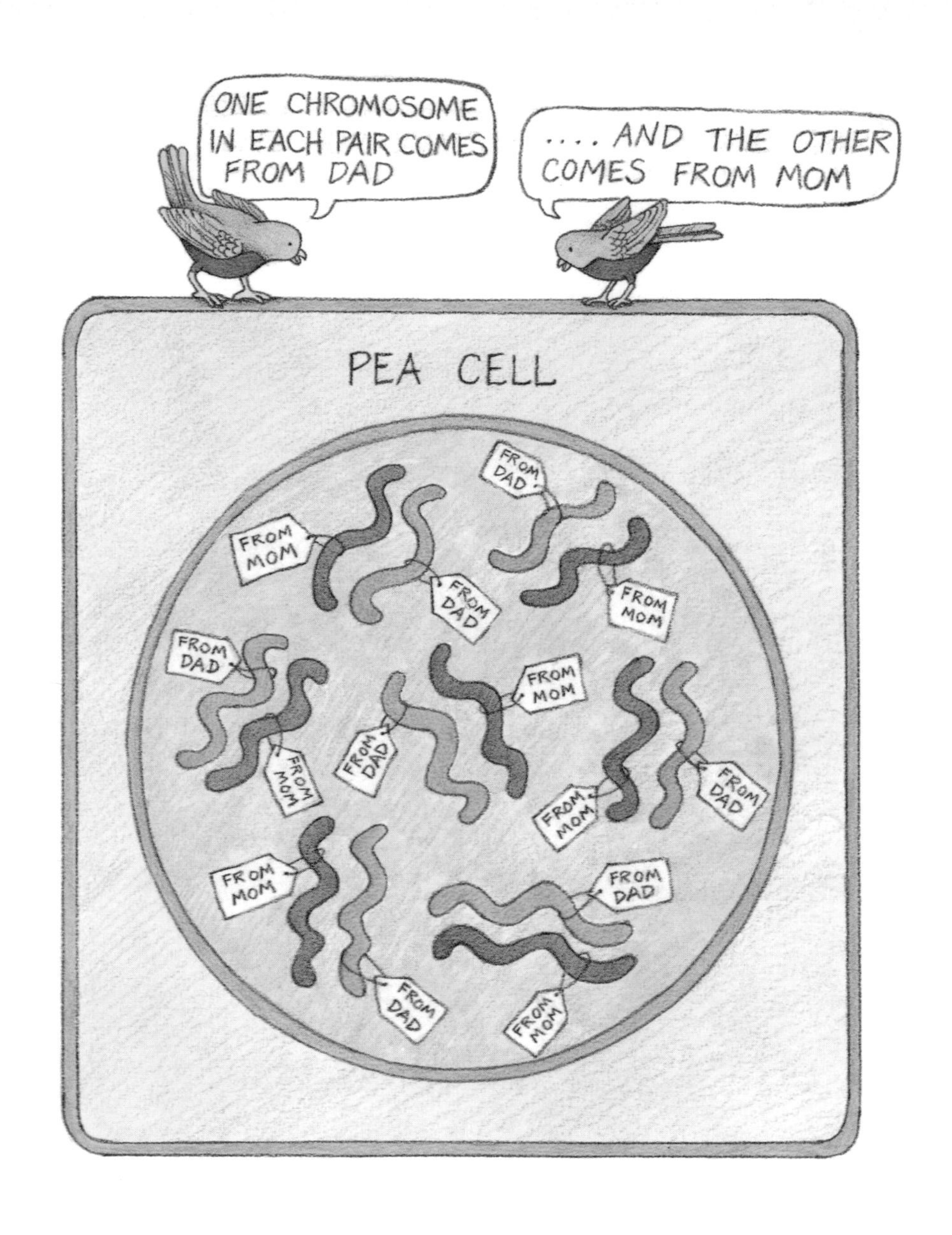

Chromosomes come in pairs. Every cell in the pea plant has 7 pairs of chromosomes, for a total of 14 chromosomes in each cell. One chromosome of each pair comes from the female parent. The other comes from the male.

Sperm and egg cells are different from the other cells in the pea plant. A pea sperm or egg cell has only 7 chromosomes, one from each pair. That way, when a sperm and egg join, the new plant or animal's cells will have the right number of chromosomes.

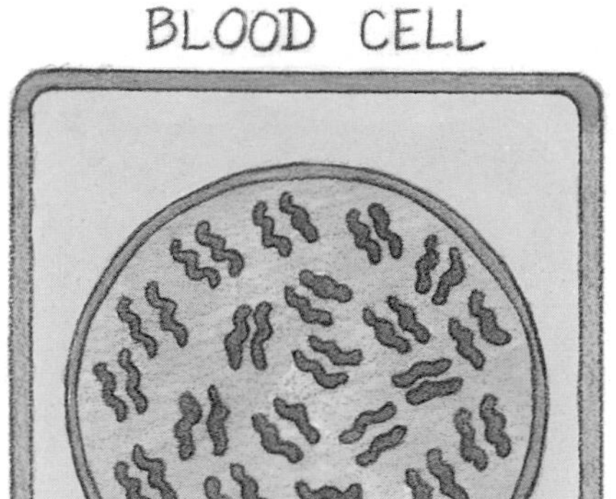

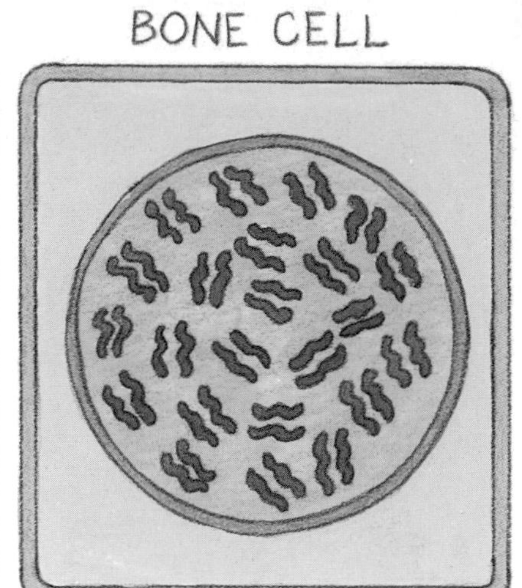

Mixing Up the Genes

People are more complicated than plants. Each of their cells has 23 pairs of chromosomes. Each one carries thousands of genes. When genes are located on the same chromosome, they are *linked.* Linked genes often go together to the eggs and sperm.

But sometimes linked genes get separated. As the cells divide, the two chromosomes in each pair come together. Before they separate to go to different cells, they sometimes trade pieces. This is called *crossing over*. Crossing over mixes up linked genes so that the same two alleles are not always found together.

Genes are made of a chemical called DNA. The DNA in each gene carries a code for making one particular protein. Proteins are very important cell chemicals. Hair is made of protein. So are muscles.

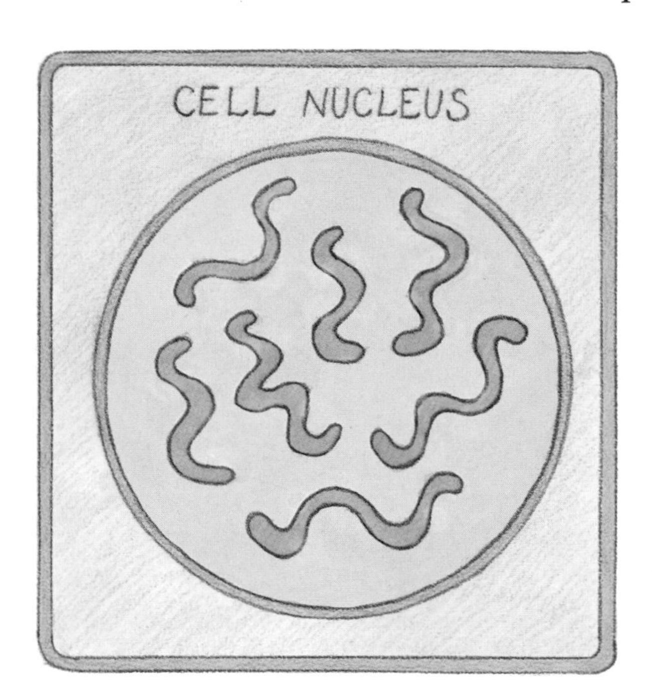

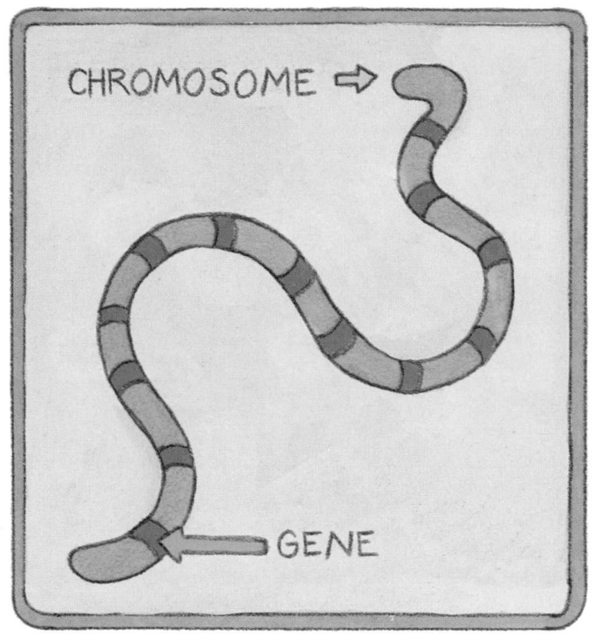

Many proteins act as *enzymes* (EN-zahyms). Enzymes control chemical reactions in cells. For example, a gene for flower color may direct the cell to make an enzyme that helps to make a red coloring.

Boy or Girl?

How do some of us get to be boys, and others, girls? Genes make the difference. One pair of chromosomes is very special. In girls and women, the two chromosomes of the pair look the same. But in boys and men, they look different. One is much shorter than the other. This shorter chromosome is called the *Y chromosome.* The longer one is the X *chromosome.*

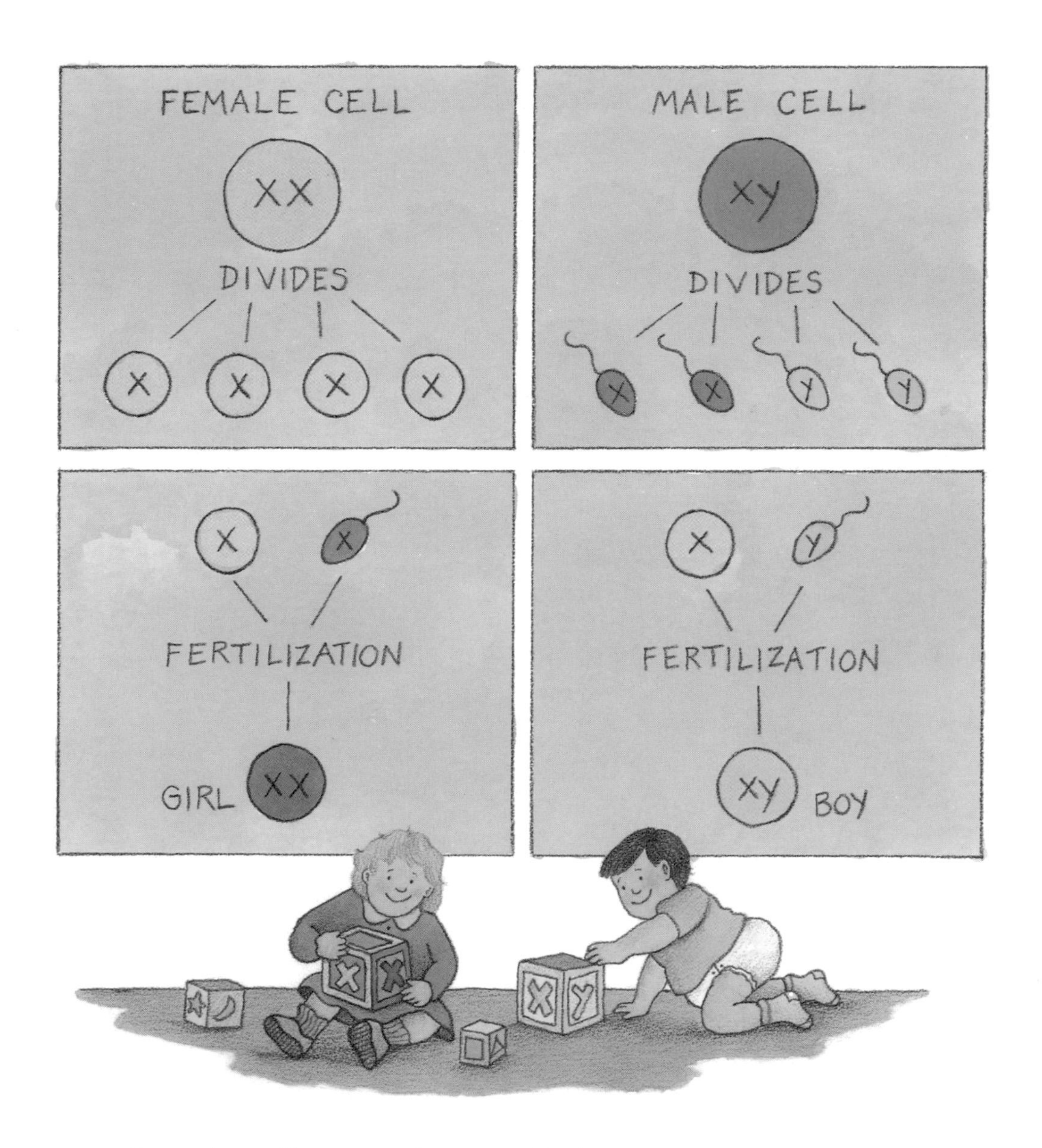

Female cells contain two X chromosomes. When cells divide to form egg cells, each egg cell will get an X chromosome. Male cells have an X and a Y chromosome. When sperm are formed, about half of them will get an X chromosome. The other half will get a Y. When a sperm with a Y chromosome fertilizes, or combines with, an egg, the baby will be a boy. When the sperm with the X chromosome fertilizes an egg, the baby will be a girl.

What Causes Differences?

Why are there different alleles of genes? Every time a cell divides to form two new cells, the chromosomes must copy themselves so that each new cell gets all the genes. Sometimes, a mistake is made during the copying. The mistake changes the gene's code, turning it into a new allele.

Mistakes can also happen during crossing over. Radiation and some chemicals can damage the chromosomes, too. All these can cause changes in the genes, called *mutations* (myu-TEY-shons). Traits like colorblindness began as mutations.

Mutations cause the differences we see among living things. Sometimes, mutations are bad. They make it more difficult to live. These mutations do not get to be common. Other mutations make it easier for an animal or plant to survive. These mutations help the plant or animal to live long enough to mature, mate (or reproduce), and have offspring that will carry the same mutation. In this way, a "good" mutation will tend to spread.

So Many Genes

Humans have about 50,000 different genes. Almost all of these have at least two alleles. Even though there are family resemblances, everyone looks different because there are so many ways the genes can be combined. This is one reason why identical twins are so remarkable.

They usually look so much alike that we can't tell them apart. Identical twins result when a fertilized egg divides into two separate cells that don't stick together. Each of the cells then goes on to develop into a complete embryo, a very young baby developing inside its mother. Both twins have exactly the same genes, since they come from the same fertilized egg. That is why they look identical.

AN EGG DIVIDING INTO TWO SEPARATE CELLS

So Many Differences

The possible combinations of genes are almost countless. Even though there are now about five billion people on Earth, no two of them look exactly alike, except for identical twins. Each human being is different from everyone else. People may say you have your grandfather's nose, but no one just like you has ever lived. And no one just like you will ever live again.

SAY
CHEESE

GLOSSARY

Alleles. Different forms of the same gene.
Chromosomes. The structures inside the cell nucleus that contain genes.
DNA. The chemical which makes up the genes.
Dominant allele. An allele which shows up even if another allele is also present.
Enzymes. Proteins that control chemical reactions.
Genes. The basic units of inheritance.
Genetics. The science that studies how traits are inherited.
Inherit. To receive the genes for traits from one's parents.
Linked genes. Genes that are on the same chromosomes.
Mendel, Gregor. The monk who discovered the laws of inheritance.
Nucleus. The part of the cell that contains chromosomes.
Proteins. The important chemicals of life. Muscles and hair are made of proteins.
Recessive allele. An allele which won't show up when paired with a dominant one.
X chromosome. One of the two sex chromosomes. A fertilized egg with two X chromosomes will become a baby girl.
Y chromosome. One of the two sex chromosomes. A fertilized egg with one X and one Y chromosome will become a baby boy.